故園畫憶

庚寅中秋
韓磬海 題

《故园画忆系列》编委会

名誉主任：韩启德

主　　任：邵　鸿

委　　员：（按姓氏笔画为序）

万　捷	王秋桂	方李莉	叶培贵
刘魁立	况　晗	严绍璗	吴为山
范贻光	范　芳	孟　白	邵　鸿
岳庆平	郑培凯	唐晓峰	曹兵武

故园画忆系列
Memory of the Old Home in Sketches

鲁西南素记
The Sketches of Southwest Shandong

黄媛媛　绘画 撰文
Sketches & Notes by Huang Yuanyuan

学苑出版社
Academy Press

图书在版编目（CIP）数据

鲁西南素记/黄媛媛绘画、撰文.—北京：学苑出版社，2015.6
（故园画忆系列）
ISBN 978-7-5077-4789-8

Ⅰ.①鲁… Ⅱ.①黄… Ⅲ.①建筑画—钢笔画—作品集—中国—现代
Ⅳ.①TU-881.2

中国版本图书馆CIP数据核字（2015）第129507号

出 版 人：	孟　白
责任编辑：	周　鼎
编　　辑：	李点点
出版发行：	学苑出版社
社　　址：	北京市丰台区南方庄2号院1号楼
邮政编码：	100079
网　　址：	www.book001.com
电子信箱：	xueyuanpress@163.com
销售电话：	010-67601101（营销部）、67603091（总编室）
经　　销：	全国新华书店
印 刷 厂：	北京信彩瑞禾印刷厂
开本尺寸：	889×1194 1/24
印　　张：	7.25
字　　数：	170千字
图　　幅：	147幅
版　　次：	2015年6月北京第1版
印　　次：	2015年6月北京第1次印刷
定　　价：	50.00元

目　录

序：文化是一种凝结于心的表达　马小依
前言

聊城市

临清清真东寺	3
临清运河钞关	4
临清大宁寺	5
聊城山陕会馆（一）	6
聊城山陕会馆（二）	7
光岳楼	8
小码头	9
铁塔	10
范公祠	11
七级镇老街	12
七级镇运河上的古闸桥	13
水月庵石塔	14
临清县治遗址	15
土桥闸（遗址）	16
陈家旧宅	17
坡里天主教堂	18
德国天主教堂	19
临清烧制贡砖技艺	20
东昌雕刻葫芦（一）	21
东昌雕刻葫芦（二）	22
东昌府年画（一）	23
东昌府年画（二）	24
阳谷哨（一）	25
阳谷哨（二）	26
聊城查拳	27
郎庄面塑（一）	28
郎庄面塑（二）	29

菏泽市

永丰塔	33
屏盗碑亭	34
巨野文庙	35
前王庄古村落	36
齐鲁会盟台	37
徐思迈妻申氏节孝坊	38
田塔	39
兴隆屯	40
百狮坊	41
佀公祠堂	42
唐塔	43
左山禅寺	44
刘忠之墓	45
朱程碑	46
肖家大院	47
付庙民居	48

朱家大院（一）	49	百年土陶老窑	78
朱家大院（二）	50	鲁南石头村落（一）	79
老地委党校	51	鲁南石头村落（二）	80
成武县老街水务局	52	葫芦套老村（一）	81
曹州面人（一）	53	葫芦套老村（二）	82
曹州面人（二）	54	葫芦套老村（三）	83
鄄城砖塑	55	孙家大院	84
鄄城旋木玩具（一）	56	电光楼	85
鄄城旋木玩具（二）	57	飞机楼	86
菏泽索弦乐	58	鼓儿词（一）	87
水浒牌纸牌（一）	59	鼓儿词（二）	88
水浒牌纸牌（二）	60	山亭区皮影	89
山东落子	61	伏里土陶（一）	90
枣梆	62	伏里土陶（二）	91
定陶皮影	63	人灯舞	92
仿山山会	64	洛房泥玩具	93
		张范剪纸（一）	94
枣庄市		张范剪纸（二）	95
甘泉禅寺（一）	67		
甘泉禅寺（二）	68	**济宁市**	
甘泉禅寺（三）	69	曲阜孔庙（一）	99
龙泉古塔	70	曲阜孔庙（二）	100
台儿庄清真寺	71	四基山观音庙	101
台儿庄古城	72	重兴塔	102
牛山孙氏宗祠	73	兴隆塔	103
报国塔	74	金口坝	104
滕州汉画像石（一）	75	青莲阁	105
滕州汉画像石（二）	76	上九山村	106
龙牙山石刻	77	古城城墙	107

柳行东寺	108	鱼台木版年画（二）	134
东大寺帮克亭	109	**临沂市**	
漕井桥	110	宝泉寺	137
苊园	111	迎仙桥	138
南阳古镇	112	蒙山千年古村落	139
清代古城门	113	马牧池乡村落	140
魁星楼	114	临沂五贤祠	141
冉子祠	115	吴白庄汉墓	142
太白楼	116	丛柏庵	143
宝相寺塔	117	地主大院	144
宝相寺鼓楼	118	戏台子	145
九龙山摩崖造像石刻	119	马牧池民居	146
凤凰山唐代佛造像	120	王庄乡天主教堂	147
嘉祥武氏墓群石刻	121	临沂天主教堂	148
嘉祥武氏墓群石狮	122	华东革命烈士陵园	149
潘家大楼	123	银雀山汉墓竹简博物馆	150
微山朝阳洞旁的民居	124	中共山东分局旧址	151
日军碉堡	125	西汉陶车陶马俑	152
山头花鼓戏	126	西汉陶俑	153
兖州花棍舞	127	小郭泥塑（一）	154
二人斗	128	小郭泥塑（二）	155
虎头袢子	129	布老虎	156
嘉祥跑竹马	130	送火神之踩高跷	157
文圣拳	131	弦子戏	158
梁山梅花拳	132	绣香包	159
鱼台木版年画（一）	133		

Contents

Preface: Culture is a Kind of Condensed
Expression in the Heart   Ma Xiaoyi
Prologue

Liaocheng City	1
Liaocheng City	3
Linqing Canal Customs	4
Linqing Daning Temple	5
Liaocheng Shanshan Guild Hall (1)	6
Liaocheng Shanshan Guild Hall (2)	7
Open Hall of Guangyue Tower	8
Small Pier	9
Iron Tower	10
Fan's Shrine	11
The Ancient Xiazha Bridge of Seven Seven Steps Ancient Town	12
The Ancient Xiazha Bridge of Steps Canal	13
Stone Tower of Shuiyue Nunnery	14
The Site of Linqing County Hall	15
Site of Bridge Sluice	16
The Former Residence of Family Chen	17
Poli Cathedral	18
German Cathedral	19
Skill of Making Bricks in Linqing	20
Dongchang Carved Gourd (1)	21
Dongchang Carved Gourd (2)	22
Dongchang New Year Pictures (1)	23
Dongchang New Year Pictures (2)	24
Yanggu Whistle (1)	25
Yanggu Whistle (2)	26
Liaocheng Zha Boxing	27
Langzhuang Dough Modelling Making (1)	28
Langzhuang Dough Modelling (2)	29
Heze City	31
Yongfeng Pagoda	33
Pingdao Stele Pavilion	34
Juye Confucius Temple	35
Ancient Village of Qianwangzhuang	36
Qilu Alliance Platform	37
Xu Simai Wife's Chastity Arch	38
Tian Tower	39
Xinglong Village	40
One Hundred Lion Archway	41
Si Ancestral Temple	42
Tang Pagoda	43
Zuoshan Zen Temple	44
Liu Zhong's Tomb	45
Zhucheng Monument	46
Xiao's Compound	47
Fumiao Dwelling House	48
Zhu's Compound (1)	49

Zhu's Compound (2)	50	Stone Village of South Shandong (1)	79
Former Party School of Heze Region	51	Stone Village of South Shandong (2)	80
Chengwu Water Supplies Bureau	52	Hulutao (Gourd) Village (1)	81
Caozhou Dough Figurine (1)	53	Hulutao (Gourd) Village (2)	82
Caozhou Dough Figurine (2)	54	Hulutao (Gourd) Village (3)	83
Juancheng Brick Molding	55	Sun's Compound	84
Juancheng Spinning Wooden Toy (1)	56	Searchlight Tower	85
Juancheng Spinning Wooden Toy (2)	57	Aircraft Tower	86
Heze String Music	58	Guerci (1)	87
Making Water Margin Playing Cards (1)	59	Guerci (2)	88
Water Margin Playing Cards (2)	60	Shanting Shadow Play	89
Shandong Laozi (Local Opera)	61	Fuli Pottery (1)	90
Zaobang Opera	62	Fuli Pottery (2)	91
Dingtao Shadow Play	63	Rendeng (Figure Light) Dancing	92
Fangshan Mountain Fair	64	Luofang Clay Toys	93
		Zhangfan Paper-cutting (1)	94
Zaozhuang City	65	Zhangfan Paper-cutting (2)	95
Ganquan Zen Temple (1)	67		
Ganquan Zen Temple (2)	68	**Jining City**	97
Ganquan Zen Temple (3)	69	Confucious Temple of Qufu (1)	99
Longquan Ancient Pagoda	70	Confucious Temple of Qufu (2)	100
Taierzhuang Mosque	71	Goddess of Mercy Temple of Sijishan	101
Taierzhuang Ancient Town	72	Chongxing Tower	102
Niushan Sun's Ancestral Hall	73	Jinkou Dam	103
The Patriotic Tower	74	Xinglong Pagoda	104
Tengzhou Han Dynasty Stone Portraits (1)	75	Qinglian Pavilion	105
Tengzhou Han Dynasty Stone Portraits (2)	76	Shangjiushan Village	106
Longyashan Carving Stone	77	Ancient Town Wall	107
100-Year-Old Earthenware Kiln	78	Jining Liuxingdong Temple	108

Bangke Pavilion of Grand East Mosque	109
Caojing Bridge	110
Jin Garden	111
Nanyang Ancient Town	112
Ancient Town Gate of the Qing Qynasty	113
Kuixing Tower	114
Ranzi Shrine	115
Taibai Restaurant	116
Baoxiang Temple Drum Tower	117
Baoxiang Temple Pagoda	118
Jiulong Mountain Cliff Stone Statues	119
Phoenix Mountain Buddha Statues in the Tang Dynasty	120
Carved Stone of Mrs. Wu's Tombs in Jiaxiang	121
Stone Lions of Mrs. Wu's Tombs in Jiaxiang	122
Pan's Compound	123
Dwelling Houses beside Chaoyang Cave in Weishan	124
The Japanese Fort	125
Shantou Flower Drum Opera	126
Yanzhou Flower Stick Dancing	127
Two People Fighting	128
Tiger Head Belt Loop	129
Riding Bamboo Horse in Jiaxiang	130
Wensheng Boxing	131
Liangshan Plum Blossom Boxing	132
Yutai New Year Woodblock Pictures (1)	133
Yutai New Year Woodblock Pictures (2)	134

Linyi City	**135**
Baoquan Temple	137
Yingxian Bridge	138
1000-Year-Old Ancient Village in Mengshan	139
Village in Mamuchi Town	140
Linyi Wuxian Shrine	141
Tomb of Han Dynasty in Wubaizhuang Villaeg	142
Cypress Nunnery	143
Landlord Compound	144
Opera Stage	145
Mamuchi Dwelling Houses	146
Wangzhuang Cathedral	147
Linyi Cathedral	148
Revolutionary Martyr Cemetery of East China	149
Yinqueshan Bamboo Slips Museum	150
Site of the Communist Party of China Committee, Shandong Branch	151
Pottery Cart and Pottery Horse of West Han Dynasty	152
Pottery Figures of West Han Dynasty	153
Xiaoguo Clay Sculpture (1)	154
Xiaoguo Clay Sculpture (2)	155
Cloth Tiger	156
Walking on Stilts to Send the God of Fire	157
Xianzi Opera	158
Sachet	159

序：文化是一种凝结于心的表达

在艺术的创作中，除去要呈现外在的艺术形象轮廓，最为重要的就是在阐释内容时形成它的思想主题。艺术之难，难于文化。难在把文化作为承载思想的依托，传达出源自"根性"的厚重感。而对于黄媛媛创作的这批古建筑、古村落以及非物质文化遗产的零散片段，我依稀看到了她对文化的执着表达和质朴情意。

传统文化离我们有多远？似乎是一个很难说清的问题，但我真的不愿把传统文化与"熟视无睹""司空见惯"联系起来。在一个城市人口密集、经济高度发展的时代，我们却依然可以看到残旧的古建筑亟待修缮，依然可以看到白发苍苍的老者一生的技艺无法传承……在一个有关物质与精神话题的探讨中，我们同时，也看到了文化人的坚守，看到了物质功利的逼迫。

其实，在我们远离传统文化的同时，它也宣告了与我们的擦肩而过。从这样一个由头说起，我觉得黄媛媛这本书的集结出版至少是一个愿意承担传统文化思想的积极表现，一个由审美心灵勾勒出的历史缩影。她的速写以"意象"捉取具有表现性的画面，所刻画的古建筑讲求形神统筹，并服从于文化的主题中。落笔之难便难在"意态"的挖掘和体现上，怎样透过建筑之美去凸显它的"意态"这是极为困难的。黄媛媛根据艺术创作法则的逐步推移、延伸，并结合美术史学的大量积累，准确重塑了古建筑的精神风韵。

黄媛媛专注于对古文化的表述，在描绘古建筑及古村落时她不求"似"，而求"真"。她不以逼真刻画物象形态为旨归，而是力求挖掘本真质朴的状态。从这个层面来说，黄媛媛对于古建筑的速写理解不是停留于模仿状态的，她不是表像化的"摹古"，而是以情感顿悟的状态切入语境，体验文化，以心状物。因而，在不断进行物、我对话的深层体验中，黄媛媛对于她所描绘的那个伫立在地缘坐标上的古建筑物象有了更为深刻明晰的解读，从而，在古建筑风貌的普遍性中抽离出她想要具体表现的它的个性特质与典型性。

在速写语言上，她注意对线条合理的把控传达内心情绪，所绘线条力避生涩、僵直之态，而是以刚中带柔、曲直相生的变化传达动态的节奏与韵律感。那种微妙的线条语言跳出了绘画语素的束缚逐步独立出来，在藏与露、破与立的对立统一中显现出自身的韵味与美感。可以说，黄媛媛在美学范畴中以中国画的笔墨线条为意象主导纳入到速写视野中，赋予速写线条以极大的自由性与灵动率意，摆脱了速写极易"刻板、呆滞"的缺憾，呈现出与传统国画相互贯通的艺术精神。在跟随画境逐步延展开来的视觉呈现中，我们看到了菏泽、济宁、枣庄等地的文化风俗与特色，这凝集着鲁西南传统文化的审美格调：朴拙、厚重、典雅、沉郁。在这种艺术气质的弥散中，我们看到鲁西南古建筑本身还有一层深意就是"历史文化的存在感"。它们不同于任何一个当代的建筑形态或样式，它们按照古人对文化审美的标准而建造，历经朝代更迭、战乱动荡、自然损毁的岁月变迁，呈示出的是一份历史与文化浇注的民族精神，是一种"大美沧桑"的文化担当。可以说，那种厚重的"文化感"是让我们心生敬畏的。但对于它的现状，我也有着这样的深思和忧虑：这种"文化感"在我们当下的艺术语境中处于怎样的状态？它是否还与我们的文化生活有着一种密不可分的心灵维系？我们是否还愿意去亲近这样的文化风景？

其实，文化的根性是民族的灵魂，是我们区别于其他国家独有的文化内蕴，在文化差异中只有古老的传统能带领我们走进文化的高地，去体味中国式的哲学。而这正是我们在精神上赖以生存的资本。

诗性是所有文化、艺术达到最高峰的精神领会。那种柔软细腻、含蓄内敛的诗意在古建筑坚硬的外表下藏匿着、隐喻着，迎接着每一个短暂生命的目光与之对接，又与之错过。我们的渺小，在诗意的文脉中渐次清晰：从乡间到地头，从城市到山区，每一串足迹，每一个片段，从独立到连结再到交叠，融汇其间的是我们与先人智慧的相知相遇，是我们短暂回眸与文化今生结缘的深情烙印。

2014 年 11 月

Preface:Culture is a Kind of Condensed Expression in the Heart

In the creation of art, besides presenting external artistic image contour, the most important component is to form the theme while elaborating on the content. The key and difficult points of artistic expression lie in using culture as a carrier of thought and conveying the in-depth insight of history derived from the "root."

It is hard to measure how far away we are from our traditional culture. In an era with densely populated cities and a highly developed economy, there are ancient buildings urgently needing to be repaired and life-long skills from white-haired old men that need to be inherited. In the contest between material and spirit, we see the insistence of intellectuals as well as the persecution of material utility, at same time.

The publishing of Huang Yuanyuan's works is at least a positive sign and shows her willingness to undertake the traditional culture, thoughts and historic microcosm outlined by an aesthetic heart. Her sketches catch the expressive moments of images and the ancient architecture she draws has integrated the shapes and spirits in harmony, as well as expressed the cultural theme.

Huang Yuanyuan focuses on expressing ancient culture while striving to find a natural pristine state. To this end, Huang Yuanyuan not only draws sketches of ancient buildings by imitation, but also gives insight into the culture while infusing her drawings with heart.

In the process of sketching, she controls the lines properly to convey emotions. She conveys dynamic rhythm through her sometimes soft, sometimes solid, straight, and at other times, curled lines. She gives her lines great freedom and a sense of spirituality. We see the folk culture of Heze, Jining and Zaozhuang through the combined dancing lines. This agglutinates the aesthetic style of the traditional culture of Southwest Shandong, which is simple but solid, elegant and gloomy. The architecture of Southwest Shandong itself is a kind of culture – it reflects the aesthetic culture and artistic pursuits of ancient people. After experiencing the change of dynasties, wars, and natural disasters, the architecture presented a kind of national spirit molded by history and culture. This legacy is both admirable and awesome. But their present situation also makes us reflect and worry about what position they fulfill in our contemporary art and how much of a link they still have to our present cultural life.

The root of culture is the soul of the nation, and our unique cultural connotation makes us different from other countries. Of the cultural differences, only the old traditions can lead us to the highland of culture to appreciate the Chinese philosophy. This is what we depend on to live spiritually. From the countryside to the field, from cities to the mountains, the remnants of ancient buildings and ruins are our indelible cultural brand.

Ma Xiaoyi
November, 2014

前　言

其实，我一直有一个愿望，那就是走出城市，在乡间田垄上亲近土地，亲近那些远离我们视线的古朴建筑和民俗。那种真实感与文化紧紧地交融，我可以感觉到我们的存在是与文化不可分割的，是有根脉的，有归宿的。

这次有幸参与的学苑出版社的故园画忆项目，是一个表达我内心文化诉求的契机，借由这个机会我可以将那些古建筑、民俗之美用自己的视角进行解读。也可以说，古朴沧桑的历史对我是一种触动，是一种心灵的净化与提升，在一个与城市发展充满反差的文化环境中，我感受到了这种文化气质的沉静与肃穆，也顿悟了"默默存在"本身就展现着文化的张力。

我希望自己这本《鲁西南素记》是一个对古文化恰当的阐述和表达。速写语言是自然的、亲切的，可以在变换的线条间不断通过内心体验去触摸艺术本质的精神深意。我真心希望每一个笔触里都传达着我的渴望、我的审视与反思。在这本书中，我直观体会着历史文化演变的进程，也表达着作为一个亲历者的文化愿望：那即是文化是抵达我们内心的心灵之火，指引我们前进的精神方向，只有文化可以将我们存在的意义变的深刻，变得肯定，变得与众不同。由此看来，我努力探寻的那种对文化进行速写描绘的愿景至少成为一种这样的可能：以艺术语言的切入，捕捉驻足心灵的风景。

具体来看，《鲁西南素记》这本书主要绘制的是鲁西南地区的古建筑和风俗、文化。鲁西南地区可以说是汉、回杂居，所以其中包含了民居、佛教建筑、伊斯兰建筑、基督教建筑和民俗民风。我在尊重古建筑、风俗原貌的前提下，对山东的历史文化、风土民情进行了一次完整的梳理，以速写游记的形式勾勒鲁西南地区的文化轮廓，再现历史文化特色。

当然，由于自己的艺术语言不够纯熟以及绘画水平的拘囿，加之时间仓促，疏误之处在所难免，敬请读者批评指正。

Prologue

I have always had one wish, which is to be out of the city and close to the land, on the edge of a field. I want to be close to the primitive, simple architecture and folk customs which are far away from our sight. It seems that in this way, I can feel that our existence is inseparable with our culture, and I find that there is the root and home of our existence.

It is my honor to participate in the project of *Memory of the Old Home in Sketches* organized by Academy Press. It gives me an opportunity to express the cultural demands from my heart. With this opportunity, I can interpret the beauty of ancient buildings and folk customs with my own perspective. In a cultural environment that is in such contrast to the development of the city today, I feel that those ancient buildings and folk customs are quiet, solemn, and respectful, and they show the tension of culture.

The Sketches of Southwest Shandong is a proper elaboration and expression of ancient culture. In this book, I intuitively experience the historical process of cultural evolution and try to describe and express the heart's cry through the culture sketch. That is the spirit direction by which our culture guides us. Only culture can elucidate the deep meaning of our existence and give us the confidence to be different.

The Sketches of Southwest Shandong mainly contains drawings of ancient buildings and the customs of Southwest Shandong. In Southwest Shandong, Han people and Hui minority people live together. So, there are local-style dwelling houses, Buddhist architecture, Islamic architecture, Christian architecture and different customs in this region. Under the premise of respecting ancient buildings and original customs, I sketched the outline of the culture of Southwest Shandong to reveal its historical and cultural characteristics.

聊城市
Liaocheng City

临清清真东寺

临清三大名寺之一。始建于明成化元年（1465年），位于临清市城区卫运河东岸，有大门、二门、穿厅、正殿、南北讲经堂、沐浴室等建筑。

Liaocheng City

Built in 1465, located at the east bank of the Wei Canal, Linqing city, the temple consists of a grand gate, second gate, connection hall, main grand hall, north and south lecture hall and bathing room, etc.

| 临清运河钞关 |

位于临清市城区内运河旁。是明清两朝设于运河督理漕运税收的直属机构，也是目前现存的一处重要运河钞关遗址。

Linqing Canal Customs

Located beside the canal of Linqing City, LCC was affiliated with the central government in the Ming (1368-1644) and Qing (1644-1911) dynasties as a tax bureau for water transport. It is an important existing site of canal customs.

> 临清大宁寺

山东省较为有名的佛教寺院。位于卫运河东岸，始建年代无考，明万历（1573~1620年）、清乾隆（1736~1795年）年间先后两次重修。

Linqing Daning Temple

Located at the east bank of the Wei Canal, it is unknown when the temple was built. It was restored during the Ming (1368-1644) and Qing (1644-1911) dynasties. It is a famous Buddhist temple in Shandong province.

聊城山陕会馆（一）

由山西、陕西的商人为"祀神明而联桑梓"集资兴建。位于古运河西岸。始建于清乾隆八年（1743年）。图为会馆内的戏楼，坐东面西，为二重檐两层台楼。

Liaocheng Shanshan Guild Hall (1)

Located at the south part of the city in the west bank of the ancient canal, the Guild Hall was built in 1743 with funds from merchants of Shanxi and Shaanxi Province as a place where sacrifices could be offered. The drawing shows the opera stage of the Guild Hall, west-facing and opposite to the Grand Hall.

聊城山陕会馆（二）

山陕会馆的房脊饰物，雕刻精妙绝伦，绘画精湛高超，在国内实属罕见。

Liaocheng Shanshan Guild Hall (2)

The original ridge decoration of the Shanshan Guild Hall was exquisite. Its display of excellent skill in painting is a rarity in the country.

> 光岳楼

　　又称余木楼、鼓楼、东昌楼。位于聊城市东昌府区古城中央，始建于明洪武七年（1374年）。此图表现的是楼内的敞轩。

Open Hall of Guangyue Tower

Guangyue Tower, also known as "Yumu Tower" "Drum Tower" or "Dongchang Tower" is located at the center of the ancient town, Dongchangfu District, Liaocheng City. It was built in 1374. The drawing shows the Open Hall inside it.

小码头

位于聊城市运河大码头以北300米处的古运河北岸,东西长14米,南北宽三米,平面呈"凹"字形。原为富商私用码头,明清时期作为漕运码头。台阶最上层青石上仍有系船缆绳用的圆形穿孔。

Small Pier

The pier is located at the north bank of the ancient canal, which was 300 meters north of the grand pier of Liaocheng City. It was originally a private pier for rich businessmen. It was used for transportation during the Ming (1368-1644) and Qing (1644-1911) dynasties. There is still a round hole for mooring ropes on the top step of the stairs.

> 铁塔

聊城现存最古老的建筑。位于东关运河西岸，初建于北宋早期。为八角形楼阁式佛塔，现为12层，塔高15.8米，由塔身、塔座两部分组成。

Iron Tower

Located at the west bank of the Dongguan Canal, eastern Liaocheng City, it was built during the early Northern Song Dynasty (960-1127) as an octagonal pavilion-style pagoda. It is the oldest existing building in Liaocheng City.

范公祠

位于梁水镇,1941年4月为纪念在聊城保卫战中牺牲的范筑先老将军及700余名抗日英烈而建。其院落正中建有十几米高的"山东省第六区抗日英烈纪念塔",此塔现在仍保存完好。

Fan's Shrine

In April 1941, to commemorate the old General Fan Zhuxian and more than seven hundred other anti-Japanese heroes who died in Liaocheng Battle, the anti-Japanese army of Southwest Shandong built Fan's Temple in Liangshui Town, and it is still intact.

七级镇老街

七级镇是千年古镇，明清在此设仓转漕，因运河边修有七级石阶大码头而得名。街市铺面相连，颇为繁盛，清末河道废弃，渐失繁荣。现存的七级老街是鲁运河沿线中保存完好的商业老街之一，建筑形式基本保持清末民国的形制，部分店面前设有砖砌外廊。现为阳谷县七级镇政府所在地。

Seven Steps Ancient Town

Seven Steps Town is a 1000-year-old town. During the Ming (1368-1644) and Qing (1644-1911) dynasties, the government set up a warehouse where goods would be stored until transfer. It was named Seven Steps Town after the seven step pier of the great canal. It was very prosperous and the shops in the street were numerous. During the late Qing Dynasty (1644-1911), the water channel was abandoned and it lost its prosperity. The existing old Seven Steps Street is one of the well-preserved commercial streets in the Lu (Shandong) canal area.

七级镇运河上的古闸桥

建于元大德元年（1297年），南北长约近百米，东西宽约80米。现在虽然古闸模样仍在，但已改建成一座石桥。

The Ancient Xiazha Bridge of Seven Steps Canal

Built in 1297, the bridge was about 100-meters long from the north to the south and 80-meters wide from the east to the west. It is a stone bridge now.

水月庵石塔

位于高唐县涸河乡岳堂村东南。清乾隆年间（1736~1795年）所立。塔上记载清康熙（1662~1722年）、雍正（1723~1735年）、乾隆（1736~1795年）等年间在高唐县水月庵、白衣庵等处修行僧人的事迹。塔高3.6米，分九层。造型优美，制作精细，堪称艺术珍品。

Stone Tower of Shuiyue Nunnery

With a history of more than two hundred years, there were written records about the monks who stayed at Shuiyue Nunnery and Baiyi Nunnery of Gaotang County during the Kangxi (1662-1722), Yongzheng (1723-1735) and Qianlong (1736-1795) periods of the Qing Dynasty (1644-1911). The tower is 3.6 meters high with 9 floors.

临清县治遗址

明初临清县治所，又称文昌阁。位于临清市区考棚街西侧，始建于明洪武二年（1369年）。歇山卷棚顶，抬梁式木构梁架，筒瓦覆顶，飞檐挑角，整体结构巧妙，和谐得体。现门楣上镌刻有"县治遗址"四字。

The Site of Linqing County Hall

Located at the western side of Kaopeng Street, Linqing City, the hall was built in 1369 as the County Hall of Linqing County during the early Ming Dynasty (1368-1644) and was also known as Wenchang Tower. Four Chinese characters meaning "Site of County Hall" were engraved on lintels.

土桥闸（遗址）

明代京杭大运河上的重要水利设施，建于明成化七年（1471年）。由青石堆砌而成，包括闸门、墩台、东侧的月河及运河两岸的进水闸、减水闸等设施。现存基本保存完整。

Site of Bridge Sluice

Located on the Beijing-Hangzhou Grand Canal, built in 1471, it was one of the important water conservancy facilities on the canal. Constructed with bluestone, the structure is still complete.

陈家旧宅

陈氏家族的宅院，共五院一园八门户，安排严紧、布局得当。建于清康熙四年（1665年），清末战乱，家族主要成员相继外出他乡定居，宅院多年失修。现仅存阁楼一座，瓦房九间。

The Former Residence of Family Chen

Built in 1665, with 5 yards, 1 garden and 8 houses. Later destroyed in the war, it was abandoned for many years. Now only 1 attic and 9 rooms remain.

坡里天主教堂

位于阳谷县城北 17 千米定水镇坡里庄。是天主教阳谷教区（成立于 1933 年）的主教座堂，也是山东省西部最大的天主教堂。

Poli Cathedral

Located at Poli Village, Dingshui Town, 17 kilometers north of Yanggu County, Shandong Province. It was built as the cathedral for the Yanggu parish, which was founded in 1933; it was also the largest cathedral in the western part of Shandong Province.

> 德国天主教堂

位于阳谷县旧城中央十字街头狮子楼斜对面，是德国人1927年修建的天主教堂。

German Cathedral

Located at the central cross street, opposite to Lion Tower, Old Town of Yanggu County, it was built in 1927 by Germans.

| 临清烧制贡砖技艺 |

　　古老的汉族手工技艺，始于明永乐（1403~1424 年）初期，有 500 多年的发展历史。临清砖又名贡砖，砖窑属官办，称"皇窑"，当时曾在临清设"工部营缮分司"督烧。砖窑数百座分布在长约六七十里的运河沿岸。图为砖窑现场。

Skill of Making Bricks in Linqing

As an old Chinese craft, it has a history of more than 500 years. Hundreds of brick kilns were set up along Linqing Canal in the Ming Dynasty (1368-1644). The kiln was called "Imperial Kiln" because it was owned by the imperial court. The Linqing brick was also known as "tribute brick." The drawing shows the kiln site.

东昌雕刻葫芦（一）

古老的汉族传统手工艺品。历经千年的传承与演变，依然保持着独特的民族、地区特色和艺术风格。

Dongchang Carved Gourd (1)

Old Chinese traditional handicrafts. After about a thousand years of inheritance and evolution, it still maintains its distinctive national and regional characteristics and an artistic style.

东昌雕刻葫芦（二）

也称蚰子葫芦。据史料记载，明清时期，东昌府商贾云集，极其繁盛，当时的雕刻葫芦曾兴盛一时。

Dongchang Carved Gourd (2)

Also called Youzi Gourd. According to historical records, gourd carving was flourishing during the Ming (1368-1644) and Qing (1644-1911) dynasties.

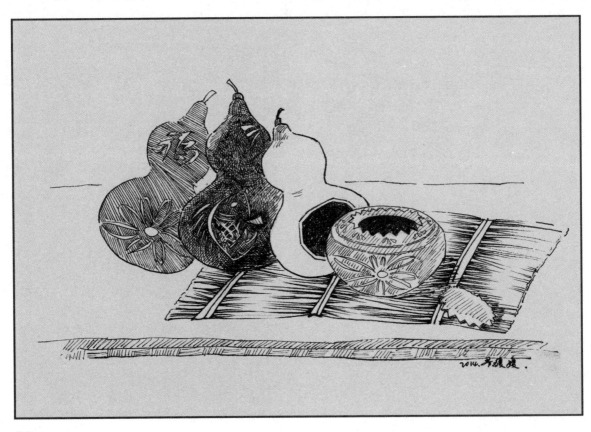

东昌府年画（一）

古代东昌府与潍坊杨家埠，并称山东两大民间画市，代表着山东木版年画的东西两大系统。很早就在国内享有盛誉，生产已有近300年历史。

Dongchang New Year Pictures (1)

Ancient Dongchang and Yangjiabu of Weifang were regarded as the two largest folk painting markets in Shandong Province. They represented the eastern and western styles of woodcut new year pictures in Shandong and were well-renowned. The history of Dongchang-style new year woodcuttings is nearly 300 years old.

东昌府年画（二）

人物形象刻画生动，悠长的线条，绚丽的颜色搭配使画面更加富有民族特色。

Dongchang New Year Pictures (2)

Vivid characters, long lines and gorgeous colors make the picture's folk characteristics richer.

阳谷哨（一）

　　流传于阳谷一带的民间娱乐用品，既可以模仿各种鸟类的叫声，又可以吹奏多种民间小调。来源于古老的乐器种类"埙"，原名"咕咕虫"，用胶泥烧制而成。

Yanggu Whistle (1)

Derived from the ancient musical instrument Xun (Chinese Ocarina), the whistle was a kind of popular home-made entertainment device. This one was created by Li Baozheng, a farmer of Danianguo Village, Yanggu County, Liaocheng City, and has a history of more than 60 years.

阳谷哨（二）

20世纪50年代聊城市阳谷县大碾郭村农民李保正发明创制，距今有60余年的历史。分为泥制、陶制两种，从三寸许至一尺大小不等。上有10孔，音域在10度以上，发音清脆嘹亮。

Yanggu Whistle (2)

Yanggu whistles can be made using either mud or ceramic and are usually about 10 to 30 cm in length. They have 10 holes with a range of over 10 degrees and produce a clear tone. These whistles can mimic the songs and calls of all kinds of birds while playing a variety of folk songs.

聊城查拳

又称回回拳，唐代查密尔、滑宗岐所创，起源于今山东冠县。在上千年的岁月里在穆斯林民众中代代相传。

Liaocheng Zha Boxing

Also called Huihui boxing, this sport originated from Guan County of Shandong Province. The tradition has been handed down from generation to generation in Muslim communities for over a thousand years.

〔郎庄面塑（一）〕

　　汉族传统手工艺品，起源于山东冠县北馆陶镇郎庄村。面塑花样丰富、题材广泛。此图是"哪吒闹海"。

Langzhuang Dough Modelling Making (1)

Made by raw flour that is kneaded, fermented, steamed, dried and colored, it was used to make various lifelike images.

郎庄面塑（二）

郎庄面塑的"绝活"在于以生面粉捏成之后并无美感，待发酵蒸熟，方见丰满。晾干、着色之后，立即形神兼备、栩栩如生。

Langzhuang Dough Modelling (2)

Originating from Langzhuang Village, Beiguantao Town, Guan County, Shandong Province, this modeling is a traditional craft treasured by the Han people with a legacy of rich patterns and various subjects.

菏泽市
Heze City

> 永丰塔

又称梵塔，位于巨野县城内，始建于唐朝。原为八角楼阁式砖塔，现存六层，高 30 余米。第一层以砖迭涩出檐，二至五层出斗拱，第六层为亭式建筑。

Yongfeng Pagoda

Also known as the Brahman Pagoda, it is located in Juye County. It was an octagonal brick pagoda built in the Tang Dynasty (618-907). The existing pagoda is more than 30 meters high with six stories.

> 屏盗碑亭

　　屏盗碑因久埋地下，免受到了人为的破坏，保存很完整。上有"大周任使君屏盗之碑""显德二年（955年）"字样。发掘出土后，移立于永丰塔西南角，并修建了碑亭。

Pingdao Stele Pavilion

Unearthed in Juye County, moved to the southwest corner of Yongfeng Pagoda and later placed inside the pavilion, the stele is well preserved as it was long buried underground and avoided corruption by human contact. Originally, there were words written on the stele.

巨野文庙

位于巨野县城东南隅、永丰塔北约 300 米处。宋金时期巨野文庙原建于城北，屡经河患，废立不一。明洪武十五年（1382年）县丞吕让移建于今址，历代多次增修扩建。

Juye Confucius Temple

Located at the southeast part of Juye County, about 300 meters north of Yongfeng Pagoda, the temple was initially built in the north part of town during the Jin Dynasty (1115-1234) and moved to the present site in 1382 during the Ming Dynasty. It has been restored and expanded many times since then.

前王庄古村落

核桃园镇前王庄村是有 500 多年历史的古村落，现有 100 多座老宅院中大部分保存基本完好。

Ancient Village of Qianwangzhuang

The Qianwangzhuang Village of Hetaoyuan Town is an ancient village with a history of more than 500 years. Of the more than 100 houses with courtyards, most are well preserved.

| 齐鲁会盟台 |

位于巨野县城西南郊大李庄村北。元末明初，由于黄河多次决口淤积，再加上雨水冲刷，台面逐渐缩小。现存台高约两米，东西长 34 米，南北宽 24 米，面积约为 816 平方米，形如覆斗状。

Qilu Alliance Platform

The platform is located at the northern part of Dalizhuang Village, southwest of Juye County. The area of the platform gradually reduced overtime due to erosion from the Yellow River and years of heavy rainfall. The existing platform is about 2 meters high and 816 square meters in area and has the shape of a covered bucket.

| 徐思迈妻申氏节孝坊 |

典型的清代石坊建筑，位于成武县张楼乡徐老家村。建于清乾隆五十二年（1787年），全石结构，造型古朴典雅、雕工考究、书法俊美。

Xu Simai Wife's Chastity Arch

Located at Xulaojia Village, Zhanglou Town, Chengwu County and built in 1787 with a structure made completely of stone, it has a primitively simplistic style and an elegant shape. In addition, it was made with exquisite carvings and beautiful calligraphy. It is typical of the stone memorial archways of the Qing Dynasty (1644-1911).

> 田塔

　　位于成武县大田集镇向东，为唐初所建石塔，该塔由塔刹、塔身、基座组成，平面呈方形，楼阁式，造型古朴端庄。

Tian Tower

Located east to Datianji Town, Chengwu County, it was built during the Tang Dynasty (618-907) as a stone tower. Its main features are a square shape in the pavilion style, a primitive simplicity, and elegant patterns.

> 兴隆屯

　　位于东明县菜园集乡黄河滩内，鲜明的住房格局颇受人关注。为躲避洪水，这一带的居民多筑台而居，这也是菏泽境内许多堌堆的起源。

Xinglong Village

This village is located at Yellow River beach in Caiyuanji Town, Dongming County. The distinct housing style made it the focus of everyone. To escape the floods, many residents here built their houses on platforms.

百狮坊

　　位于单县城内张牌坊街东端。因坊体夹柱上精雕百个石狮而得名。又称节孝坊、张家牌坊,乾隆四十三年(1778年)为文林郎张蒲妻朱氏而建。坊上百个狮子雕刻,精致并形无同。

One Hundred Lion Archway

Also called the Chastity Arch or Zhang Family Memorial Arch, it was built in 1778 during the 43rd year of Emperor Qianlong of the Qing Dynasty for Ms. Zhu, Mr. Zhang's wife. 100 lions were delicately sculpted on the archway without repeating any poses.

侣公祠堂

　　位于郓城县城东南侣楼村，建于明成化十八年（1482年）。典型的明代古建筑群，为纪念明代一代名臣侣钟（1439~1511年）所建。整体为四合院，雕梁画柱、纹兽齐全。

Si Ancestral Temple

Located at Silou Village, Southeast Yuncheng County, it was built in 1482 to commemorate Si Zhong, a famous official of the Ming Dynasty (1368-1644). It is typical of ancient building groups of the Ming Dynasty.

> 唐塔

　　位于郓城县唐塔公园内,塔下原有一座观音寺,故又称观音寺塔,原高七级。建于五代后唐长兴二年(931年),故人们又称其为"唐塔"。

Tang Pagoda

The 7-story pagoda was also known as the Goddess of Mercy Pagoda, and it was built in 931 during the Late Tang period of the Five Dynasties. This is how it received its name.

> 左山禅寺

　　位于定陶县马集镇西南。千年古刹，是我国历史上的著名寺院。隋朝名法源寺，唐改为龙兴寺，北宋又改称兴化禅院。寺院高大宏伟，古树参天，寺中藏有佛教圣物。

Zuoshan Zen Temple

Located at the southwest of Maji Town, Dingtao County, it is one of the most famous thousand-year-old temples in our country's history. The temple is grand and magnificent with old trees covering the sky and sacred Buddhist goods kept guard inside.

> 刘忠之墓

　　明弘治年间（1488~1505年）都察院右副都御史刘忠之墓，其为人正直、政绩卓著。位于菏泽市鄄城彭楼镇刘垓村西。墓地现有碑刻24座，石兽两对，山门已经按原貌修复。

Liu Zhong's Tomb

Located at the west end of Liugai Village, Gulou Town, Juancheng, Heze City, it was the tomb of Liu Zhong, an official of the Ming Dynasty (1368-1644). Liu Zhong was a righteous official with remarkable achievements. There are 24 inscriptions and one pair of stone beasts in the cemetery. The gate was rebuilt according to the design of the original one.

朱程碑

位于鲁西南革命烈士陵园内。鲁西南革命烈士陵园是1945年1月为纪念在1943年"9·27"王厂反扫荡中为国捐躯的烈士而建。时任冀鲁豫军区第五军区司令员朱程牺牲后,安葬于此。碑上刻有碑文。

Zhucheng Monument

The monument is located in the cemetery for revolutionary martyrs in the southwest of Shandong Province. Zhu Cheng, the military commander of the 5th military region of the Shanxi-Hebei-Shandong-Henan border was buried here after he died. There is an inscription on the monument.

> 肖家大院

　　肖子楚将军故居。肖子楚，菏泽市牡丹区肖老家人，官至国民党兵团副司令员。肖家大院始建于1936年，砖木结构，占地面积660平方米。

Xiao's Compound

This is the former residence of General Xiao Zichu, who was from Xiao's village in the Mudan District, Heze City. He was deputy commander of Kuomintang's army. It was built in 1936 with wood and bricks and covered an area of 660 square meters.

付庙民居

典型的明清时期建筑风格的传统民居群。位于巨野县核桃园镇付庙村，有几个合院组成。

Fumiao Dwelling House

Located at Fumiao Village, Hetaoyuan Town, Juye County, it is a traditional dwelling house group consisting of several courtyards with the architectural style typical of the Ming (1368-1644) and Qing (1644-1911) dynasties.

朱家大院（一）

具有典型的明末清初北方建筑风格的古民居。位于单县三元广场北侧，建筑风格恢弘古朴，楼房都有走廊，各小院相通。雕刻图案精美。

Zhu's Compound (1)

Located at north Sanyuan Square, Shan County, as Ming and Qing Dynasty-style buildings, Zhu's Compound has the typical northern architectural style of the late Ming Dynasty (1368-1644) and early Qing Dynasty (1644-1911). It is magnificent and its primitive simplicity is coupled with exquisite carving patterns.

朱家大院（二）

现存两楼院。西楼院正楼两层五间，中三间有走廊，青磨砖、合瓦、木柱、石基、檐下有额坊、雀替、垫板；东西厢楼结构相同，三间两层，廊檐设在底层。东楼院正楼两层三间，磨砖、合瓦，廊檐设在上层；东西厢楼结构和西楼院厢楼相同。

Zhu's Compound (2)

The existing Zhu's Compound consists of 2 houses with courtyards. The west house is 2 stories with 5 rooms in the main building and 3 rooms in the middle with a corridor. The east house is 2 stories with 3 rooms in the main building; the east and west compartments have the same structure as the west house.

> 老地委党校

位于菏泽市水建街路口，图为老地委党校的食堂。现存为厂房。

Former Party School of Heze Region
Located at the cross of Shuijian Street, Heze City, Shandong Province, it is a factory now. The drawing shows the dining hall of the former party school.

成武县老街水务局

成武县老城区街道两侧建筑，大部分为明清和民国时期所建，且多保存完好。这是其中的一部分。图为老城水务局。

Chengwu Water Supplies Bureau

Most of the buildings along the street of the old town of Chengwu County were built during the Ming (1368-1644) and Qing (1644-1911) dynasties and are well-preserved. The drawing shows the former water supplies bureau.

曹州面人（一）

民间传统的造型艺术，最早起源于位于今菏泽市牡丹区的马岭岗镇穆李村。是在古代祭天地、敬鬼神的"花供"基础上发展起来的。图为面人钟馗抓鬼。

Caozhou Dough Figurine (1)

Traditional plastic arts, initially from Muli Village, Malinggang Town, Mudan District, Heze City, were developed from the religious ceremony called "flower ceremony" for worshiping a god and ghost. The drawing shows Zhongkui catching the ghost.

曹州面人（二）

　　俗称捏面人以细小麦面和江米面为主要原料，分别加入颜料，和成不同色彩的面团。用锅蒸熟后再配上适量防腐、防蛀、防干剂等。然后借助于批刀、塑刀、小剪刀、梳子、骨簪、花纹印章等工具，用手捏制成各种各样的栩栩如生的塑像。

Caozhou Dough Figurine (2)

Using fine wheat flour and rice flour as the main raw materials, they add pigment before steaming it in a pot and adding the proper anticorrosive and moth-proofing substances. Then they knead the dough by hand with help from plastic knives, scissors, decorative seals etc.

鄄城砖塑

菏泽市特有的建筑装饰，历史悠久，保持了传统的民间捏塑和土陶工艺特色。起源于清光绪年间（1875~1908年），菏泽市鄄城县谢家砖塑，世代相传。

Juancheng Brick Molding

"Xie's Brick Molding", from Juancheng County, Heze City, has been handed down from generation to generation since the 19th century. As a unique architectural decoration of Heze City, Shandong province, it has a long history and preserves the characteristics of the traditional kneading model and earthenware craft.

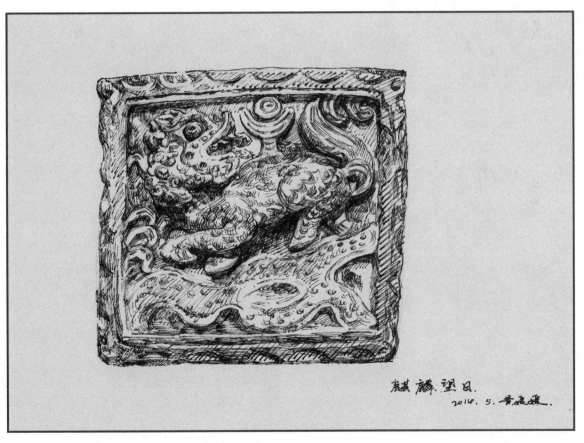

鄄城旋木玩具（一）

鄄城县郑营乡刘庄村，自古家家户户都做旋木器物，人称"旋木刘庄"。其旋木技艺，相传始于明初，代代相传，历20世而不辍。所产木玩具，从前跑河南、河北、山西、山东数省，推车挑担赶古会贩卖。

Juancheng Spinning Wooden Toy (1)

LiuZhuang Village of Zhengying Town, Juancheng County, was known as "Spinning Wood Village" because every family there has made spinning wooden toys since ancient times. Their skill in making spinning wooden toys started from the early Ming Dynasty (1368-1644) and has been handed down from generation to generation for more than 600 years.

鄄城旋木玩具（二）

手摇木玩具，当地人称"花拉棒槌"。其中央一柱类似洗衣用的棒槌，做成简单的人形并施以彩色图案。棒身内半空，装入二三粒小石子或砂子，然后安上把儿，堵住孔，摇起来哗哗作响，儿童十分喜欢。

Juancheng Spinning Wooden Toy (2)

This is a kind of hand-toy with a central column and wooden stick that is painted with different color patterns. The stick is half-empty inside with two or three small stones inside. Sometimes it is filled with sand. Install the handle and plug the hole before shaking, and it will make a swooshing noise.

| 菏泽索弦乐 |

历史悠久，风格典雅，雅乐冶遥。通常由筝、琵琶、扬琴、如意勾四种丝弦乐器演奏，也可以只用筝、扬琴二者合奏，或是筝与扬琴、琵琶合奏。主要流行于菏泽市的郓城、鄄城等县的乡村。

Heze String Music

With a long history and elegant style, Heze string music is popular in Yuncheng County and Juancheng County of Heze City. Usually played by string instruments like Zheng, Pipa, Dulcimer and Ruyi Hook, or played by two instruments like Zheng, Dulcimer and Pipa.

水浒牌纸牌（一）

起源于《水浒传》中的人物宋江的故乡郓城县水堡村。据记载，清末时水堡村有纸牌作坊100多家，为最鼎盛时期。

Making Water Margin Playing Cards (1)

Originating from Shuibao Village, Yuncheng County, Shandong Province, it was written that there were over 1000 playing card workshops at Shuibao Village during its Heyday in the late Qing Dynasty (1644-1911).

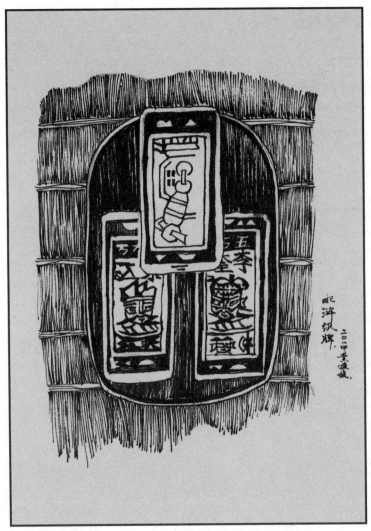

水浒牌纸牌（二）

　　水浒纸牌及纸牌游戏，最早可追溯到元代末年，已有600多年的历史。纸牌上面，绘有梁山英雄的图像，而图像的旁边，则标明当年官府捉拿梁山好汉所出的赏银数目。

Water Margin Playing Cards(2)

Dating back to the late Yuan Dynasty (1206-1368), the water margin playing cards and game have a history of more than 600 years. The portraits of 108 Liangshan heroes were painted on the cards.

山东落子

前身是"莲花落",起于宋代,盛于明清,主要流传于黄河中下游的中原地区。其唱腔与地方方言和民歌小调相结合,多以左手打单镲,右手打竹板,一人自演自唱,句末拖长腔,以鼻音结束,给人以余音绕梁的艺术冲撞力。

Shandong Laozi (Local Opera)

Formerly known as Lotus Lao, it has a history of more than 1000 years. The style of its singing combines local dialect and folk songs. An actor plays a bamboo clapper while singing and performing and ends with projecting a song in a nasally tone.

| 枣梆 |

　　流行于菏泽一带以及河北、河南省的部分地区。其渊源是约清光绪六年（1880年）前后传入山东的山西上党梆子，结合当地方言逐步演变发展而成。唱腔中真假嗓结合，真嗓吐字，假嗓拖腔。既能激昂雄壮，又能委婉动听，极具表现力。

Zaobang Opera

Popular in the Heze area and some areas of Hebei and Henan Porvince, it developed by combining the Shangdang Opera of Shanxi with local dialects and has a history of more than 100 years.

> 定陶皮影

又名隔纸说书，皮影艺术的重要地方分支。起源于清末定陶县张湾镇后冯村。相对于山东其他地方皮影的近乎说唱艺术而言，定陶皮影则近乎地方戏曲，在山东皮影中可谓独树一帜。

Dingtao Shadow Play

Also known as "paper following storyteller", Dingtao shadow play is an important local branch of shadow play art. It originated from Houfeng Village, Zhangwan Town, Dingtao County in the late Qing Dynasty (1644-1944) with a similar style to the local opera. It is a unique branch of the shadow play art of Shandong.

仿山山会

　　发源定陶县西周至春秋战国时曹国25代君王陵地，距今已有2000多年的历史。每年农历三月二十七日至二十九日是仿山山会的日子，规模不断扩大，涵盖"祭祀"和商品贸易，在其周围地区闻名遐迩。

Fangshan Mountain Fair

With a history of more than 2000 years, it has been held from March 27 to 29 on the lunar calendar every year. It has been continually expanded to an event including sacrifice and trade. Now it is well known in the surrounding area.

枣庄市
Zaozhuang City

甘泉禅寺（一）

　　位于枣庄北郊，古称伽蓝神庙，又称龙窝寺。建于元延佑六年（1319年）。现存千年树龄的银杏古树，足以佐证该寺历史之悠久。银杏树下立有明万历十五年（1588年）的石碑，题为《重修龙窝寺碑记》。

Ganquan Zen Temple (1)

Located in the northern suburbs of Zaozhuang City and built in 1319, it was known as Kuan Ti Temple or Longwo Temple in ancient times. There are one-thousand-year-old ginkgo trees in the temple that have stele from the Ming Dynasty (1368-1644) underneath them.

{ 甘泉禅寺（二）}

　　进入甘泉寺，门内西侧钟楼、鼓楼相对，各高两层。图为甘泉禅寺的钟鼓楼。

Ganquan Zen Temple (2)

Inside the Ganquan Temple, there are a bell tower and drum tower facing each other at the western side of the gate. Both towers are two stories.

甘泉禅寺（三）

　　甘泉禅寺的大雄宝殿殿前立有一座七宝石塔，制作较为精美，建造年代不详。

Ganquan Zen Temple (3)

In front of the grand Buddha's hall of Ganquan Zen Temple, there is a stone pagoda called Qibaoshi Pagoda which is very delicate. The date it was built is unknown.

龙泉古塔

位于滕州市龙泉街荆河西岸。密檐式佛塔，砖石结构，高40米，塔身呈八角九级，二挑华拱托檐，下置石砌的须弥座，前有塔室，后有塔门，周身有塔窗13个，塔顶铸铁莲花覆盆封盖，上立葫芦宝刹。

Longquan Ancient Pagoda

Located at the west bank of the Jing River, Longquan Street, Tengzhou City, the mason-constructed pagoda has dense eaves and is 40 meters high with 9 stories and an octagonal body. The entire structure of the pagoda was rigorously built.

台儿庄古城

形成于汉，发展于元，繁盛于明清。清乾隆皇帝下江南时，途经台儿庄，称之为"天下第一庄"。1938年震惊中外的台儿庄大战，使这座古城在战火中化为废墟。2008年重建。

Taierzhuang Ancient Town

This town was established during the Han Dynasty (25-220), developed in the Yuan Dynasty (1206-1368), and prosperous in the Ming (1368-1644) and Qing (1644-1911) dynasties. Qing Emperor Qianlong called it "first town in the world". The shocking Taierzhuang battle in 1938 transformed the town into ruins. It was rebuilt in 2008.

> 台儿庄清真寺

　　俗称北大寺，清乾隆七年（1742年）由阿訇李中和主持兴建，建有礼拜殿25间，大、小讲堂八间及配房、门楼、过厅等。整体建筑庄严肃穆、典雅辉煌，具有鲜明的民族特色。

Taierzhuang Mosque
Commonly known as the Grand North Mosque, it was built by imam Li Zhonghe in 1742. The whole building looks solemn and elegant with its compact layout and distinctive national features.

牛山孙氏宗祠

　　始建于明弘治元年（1488年），坐落在牛山村内。祠堂坐北朝南，两进院子，为传统的硬山式砖木结构。

Niushan Sun's Ancestral Hall

Located at Niushan Village, built in 1488, the south-facing ancestral hall has 2 rows of yards.

报国塔

位于枣庄峄城区青檀寺后的一座小山上，砖木结构，高48米，八角七级，异型灰色亚光琉璃瓦复顶，金山石浮雕护栏围绕塔廊，塔身外观为银灰色，雄伟、典雅、古朴。青檀寺是一座千年古寺，近年重建。

The Patriotic Tower

Located on the hill behind Qingtan Temple, Yicheng District, Zaozhuang City, the 48-meter-high tower is a brick structure with 7 stories in an octagonal shape. The tower is magnificent, elegant and with characteristic silver-grey color.

滕州汉画像石（一）

滕州汉画像石馆位于荆河桥畔、龙泉塔下，汉画像石西王母牛羊车，雕工精细完美。

Tengzhou Han Dynasty Stone Portraits (1)

Located beside the Jing River bridge under the Longquan Pagoda, it features Han Dynasty (25-220) stone portraits of the oxen and sheep carts of the "West Queen" in perfect, delicate carvings.

{滕州汉画像石（二）}

　　图画表现的是两个神兽交织在一起的画面，栩栩如生。雕刻严谨、细致、精美，体现了古代的雕塑工匠精湛的技艺。

Tengzhou Han Dynasty Stone Portraits (2)

The picture shows the lifelike scene of two mythical creatures intertwined. The carving is rigorous, exquisite and shows the excellent sculpting skills of the ancient craftsmen.

龙牙山石刻

　　龙牙山位于枣庄山亭区水泉镇，海拔 278 米。峭壁上的五尊石刻佛造像保存完好，为三佛二力士。天然岩洞南边的绝壁上，凿刻有"古刹宝峰禅林寺"字样。

Longyashan Carving Stone

Located at Shuiquan Town, Shangting District, Zaozhuang City, Longyashan Hill is 278 meters high with 5 stone statues of Buddha on its cliff. Among the 5 statues, 3 are of Buddha and 2 are Malla. The Chinese charactors engraved on the cliff mean "Ancient Zen Temple".

百年土陶老窑

在枣庄市市中区孟庄镇尚岩村，百年土陶老窑至今还在使用，图中摆放的是已烧制好的土陶制品。

100-Year-Old Earthenware Kiln

Located at Shangyan Village, Mengzhuang Town, Shizhong District, Zaozhuang City, the 100-year-old earthenware kiln is still in use. The drawing shows finished earthenware products.

鲁南石头村落（一）

枣庄市山亭区，零星分布着一些保存较为完整的石头村落，包括花家泉村、高山后村、邢山顶村、高山顶村、米山顶村、兴隆庄等。房屋的主体是由一块块硕大的石块堆砌而成，而屋顶则是由一片片石片构成。

Stone Village of South Shandong (1)

There are some well preserved stone villages scattered in the Shanting District of Zaozhuang City. These stone villages are called the slate house villages of South Shandong. The main bodies of houses are made by piling large stones while the roofs are made up of pieces of stone chips.

鲁南石头村落（二）

高山后村的石头房子建于清乾隆年间（1736~1795年）至新中国成立初期，村民因地制宜，就地取材，建造石板房居住。石质的围墙将每家每户分隔的井井有条，炮楼庄严地矗立在村子的各个角落。

Stone Village of South Shandong (2)

The stone houses of Gaoshanhou Village were built from the Qianlong period (1736-1795) of the Qing Dynasty to the beginning of the founding of modern China. The villagers took stones from nearby to build stone houses for living.

葫芦套老村（一）

位于薛城区北 6000 米的蜡山山腰处，现有新旧院落 112 处，大小房屋 446 间。整个村落基本上保持了晚清鲁南民居的风格。

Hulutao (Gourd)Village (1)

Located halfway up Lashan Mountain, which is 6000 meters to the north of the Xuecheng District, there are 112 new and old houses and courtyards with a total of 446 rooms of different sizes. The whole village basically maintained the dwelling house style of south Shandong in the late Qing Dynasty (1644-1911).

葫芦套老村（二）

　　始建于清乾隆二十五年（1760年），刘氏始祖刘琚琏由柴胡店镇刘村迁此建村。因村四周山峦环绕，状似葫芦，故取村名葫芦套。

Hulutao (Gourd) Village (2)

It was built in 1760 by Julian Liu, the earliest ancestor of the Liu family, who moved here from Chaihudian Town. The village was surrounded by mountains and looked like a gourd. It was named for this reason.

葫芦套老村（三）

一代又一代的村民就地取材，利用山上的青石垒建成具有鲁南特色的石板房。

Hulutao (Gourd) Village (3)

From generation to generation, villagers used the bluestones from the mountain to build stone houses with south Shandong characteristics and style.

> 孙家大院

位于枣庄市薛城北的原夏庄乡西仓村内,据考证为明末清初孙氏先人建造,距今已近400年,现村里人仍习惯称其为"孙家大院"。

Sun's Compound

The compound is located at Xicang Village, Yuanxiazhuang Town, North of Xuecheng, Zaozhuang City. According to research, it was originally built by an ancestor of Sun's family with a history of over 400 years. Villagers called it Sun's Compound.

> 电光楼

位于原枣庄煤矿八大家宿舍的东北角，是20世纪30年代日本侵略者修建的一座碉堡（炮楼）。因楼顶装有探照灯，故被人们称为"电光楼"。

Searchlight Tower

Situated at the northeast side of the dormitory of the former Zaozhuang coal mine, it was a fort built in the 1930's by Japanese invaders. There were searchlights on the roof and it was named after this.

85

> 飞机楼

　　位于枣庄市中心街，始建于1923年，原为中兴煤矿公司办公楼。因其外形酷似飞机，故又名"飞机楼"。

Aircraft Tower

Located at the central street of Zaozhuang City, built in 1923, it was originally used as the office building of the Zhongxing coal mine. It was known as Aircraft Tower because of its shape.

鼓儿词（一）

汉族曲艺说唱剧种，又称枣庄小鼓、石门小鼓。起源于枣庄市市中区一带，最早可追溯至明末清初，已有400多年的历史。明末进士石元郎是鼓儿词的始祖。流传于鲁南、鲁西南和苏北地区。

Guerci (1)

Also known as Zaozhuang Small Drum or Shimen Small Drum. It originated from the downtown area of Zaozhuang. It was a kind of old, unique and rare rap-opera of the Han people, which is popular in south and southwest Shandong, as well as northern Jiangsu province.

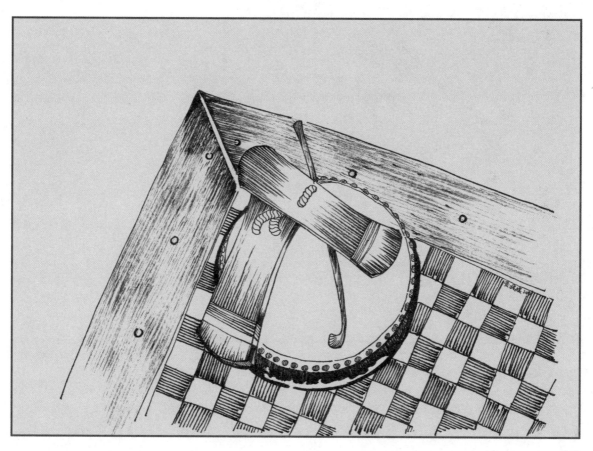

鼓儿词（二）

男或女一人表演，演唱者左手持板，右手执鼓棒，自敲自击自唱，用的是鲁南方言。唱词简练、风趣、幽默。

Guerci (2)

It's a kind of rap-opera performed by a single actor or actress. With a board in the left hand and a stick in the right, the singer sang in dialect while beating the board. The lyrics usually were witty and humorous.

山亭区皮影

民间剧种，其原生态地集民间音乐、戏曲、曲艺、手工艺、画绘于一身。唱词、道白具有鲁南民俗语言特点。

Shanting Shadow Play
A collection of folk music from traditional Chinese opera, Chinese folk art forms, arts and crafts, and paintings. The lyrics and spoken parts were full of the folk language features of South Shandong.

伏里土陶（一）

　　枣庄市西集镇伏里村的制陶业有着6000多年的悠久历史，人们在日常生活和各种民俗活动中一直使用民间艺人们自行烧制的各种红陶、黑陶、灰陶、白陶器皿。

Fuli Pottery (1)

With a long history of 6000 years, the pottery industry in Fuli Village, Xiji Town, Zaozhuang City, fires a variety of red pottery, black pottery, grey pottery and white pottery household utensils, which are widely used in people's daily lives and folk activities.

伏里土陶（二）

土陶艺术品中最具代表性的作品是蟾蜍，蟾蜍又名"避邪"。

Fuli Pottery (2)

The toad was the most representative image of the pottery crafts. Toad was also known as Avoiding Evil Spirits as people believed that it had ability to ward off evil spirits.

人灯舞

　　流行于鲁南地区的民间文化活动。起源于薛滕一带，以薛城为主要活动区域。形成于明末清初，俗称"人灯"。一般在盛大节日如农历正月十五"元宵节"举行。

Rendeng (Figure Light) Dancing

Popular in south Shandong, it originated from the Xueteng region and was mainly performed in the Xuecheng region. Formed in the late Ming Dynasty (1368-1644), it was commonly known as "Figure Light". It was usually performed during grand festivals, such as the Lantern Festival on Jan. 15th of the lunar calendar.

洛房泥玩具

距今有近 200 年的发展历史，形象分人物和动物两大类。用当地的一种白糖土制作的，此土细腻、黏合，制作出的产品不但不破裂，而且还光洁。

Luofang Clay Toys

The toys were made by a kind of local sugar soil which is exquisite and binds well. Products made by this soil are not only solid, but also bright and clean. The clay toys include figures and animals and have a history of nearly 200 years.

张范剪纸（一）

　　剪纸是一种民间艺术，见证了民间习俗，陶冶民情民风。枣庄市张范镇的剪纸已有 500 多年历史。

Zhangfan Paper-cutting (1)

Paper-cutting is a kind of folk art which reflected and was cultivated from folk customs. Zhangfan paper-cutting has a history of 500 years.

| 张范剪纸（二） |

张范剪纸工艺渊远流长。全镇剪纸能手众多，作品取材广泛，造型夸张、神采飞扬。

Zhangfan Paper-cutting (2)
Zhangfan paper-cutting has a long history and many people in this town are good at it. Their works have various subjects with exaggerated and energetic modeling.

济宁市
Jining City

曲阜孔庙（一）

位于曲阜市的孔庙棂星门，是孔庙的第一道大门，是文庙中轴线上的牌楼式木质或石质建筑，象征着孔子可与天上施行教化、广育英才的天镇星相比。始建于明永乐十三年（1415年），清曾重修扩建。

Confucious Temple of Qufu (1)

Located in the Confucius Temple of Qufu, LingXing Gate was the first gate of the temple. It was wood or stone architecture originally with a decorated archway on the axis of the temple and stands for the good virtues and high status of Confucius. It was built in 1415 and expanded in the Qing Dynasty (1644-1911).

曲阜孔庙（二）

　　重光门为孔府各建筑中的最高规制。又称仪门、塞门，始建于明弘治十六年（1503年），因门上悬明嘉靖皇帝朱厚熜御赐"恩赐重光"匾额而得名。古时平时关闭，只在皇帝驾临、迎接圣旨、祭孔、婚丧等大典活动中才可打开。

Confucius Temple of Qufu (2)

Also known as Yi Gate or Sai Gate and built in 1503, Chongguang Gate was named after the written plaque "En Ci Chongguang" ,which means commending notable family, from the emperor during the Ming Dynasty (1368-1644) . It only opens for sacrificial activities and is regarded as the top level building among Confucius architecture.

四基山观音庙

位于曲阜市尼山镇大烟庄村南四基山北麓,始建于明代,清代多次进行维修。主体建筑已损坏多年,建筑群基本垮塌倾倒,只存残垣断壁。

Goddess of Mercy Temple of Sijishan

Located at the northern foot of Sijishan Mountain, south of Dayanzhuang Village, Nishan Town, it was built in the Ming Dynasty(1368-1644) and repaired many times during the Qing Dynasty (1644-1911). The main building was damaged years ago and only ruins remain today.

重兴塔

建于宋代。位于邹城市旧汽车站西南,古塔小区院内。

Chongxing Tower

Located at the Ancient Tower neighbourhood, southwest of a former bus station, it was built during the Song Dynasty (960-1279).

| 兴隆塔 |

空心砖塔，位于兖州城内东北隅今博物馆院内，因该塔坐落在兴隆寺内，故命名为兴隆宝塔。建于隋仁寿二年（602年），距今已有1000多年的历史。塔身高54米，八角楼阁式，13层。

Xinglong Pagoda

Located in a museum courtyard in the northeast corner of Yanzhou City, the pagoda was built in 602. The tower was located in the Xinglong Temple, for which it was named.

> 金口坝

水利枢纽。始建于北魏，横跨距兖州城东 2500 米的泗河上。有着近 1500 年历史，素有"江北都江堰"之称。

Jinkou Dam

Built during the Northern Wei Dynasty (386-534) and crossing the Sihe River, which is 2500 meters to the east of Yanzhou City, it is a water conservancy hub with a history of nearly 1500 years. It is also known as "Dujian Dam at North Yangze River".

青莲阁

位于兖州市城东金口坝北侧泗河西岸，为纪念李白而建。始建年代已无可考，现仅存的只有这三间二层砖木结构的建筑。

Qinglian Pavilion

Located at the west bank of the Sihe River, north of Jinkou Dam, east of Yanzhou City, the pavilion was built in memory of Li Bai and no research has shown when the Pavilion was actually built. Now only these 3, 2-story brick buildings remain.

上九山村

位于石墙镇驻地西南8000米，村落呈方形，主街不成规则。始建于北宋初年，明洪武年间（1368~1398年）郑、聂、满三氏由山西迁来定居，因周围有大小九个山头，故取名为上九山村。

Shangjiushan Village

Located at 8000 meters to the southwest of Shiqiang Town, the village was in the shape of a square and the main streets didn't have a regular layout. Built in the early Northern Song Dynasty (960-1127), Zheng, Nie and Man families moved here from Shanxi during the Hongwu period (1368-1398) of the Ming Dynasty. There are nine mountains surrounding it.

古城城墙

济宁古城的重要象征，见证了济宁历史的发展变化。墙现残存126米，位于城区环城北路与环城西路交界处。古城墙一共经西汉早期、金代和明代三个历史时期建成。明洪武三年（1370年）重修时"易土为砖"。

Ancient Town Wall

It's one of the most important symbols of the ancient Jining City. Located at the junction of Huancheng North Road and Huancheng West Road, the existing wall is 126 meters long and was constructed through the early West Han (205 B.C,-25 A.D.), Jin (1115-1234) and Ming (1644-1911) dynasties.

> 柳行东寺

　　位于济宁市柳行南街路西，始建于明万历年间（1573~1620年）。随着漕运发达，回族人口增加，经济富裕，于清康熙年间（1622~1722年）按西大寺式样扩建。

Jining Liuxingdong Temple

Located west of Liuxing South Road, Jining City, it was built during the Wanli period (1573-1573) of the Ming Dynasty. With the development of water transportation, the flourishing economy, and the increased population of the Hui minority people, it expanded during the Kangxi period (1622-1722) of the Qing Dynasty.

东大寺帮克亭

位于济宁市小闸口上河西街。始建于明洪武年间（1368~1398年），后经明、清各朝及当代数次修缮，院内迎面是"帮克亭"。"帮克"是波斯语音，译意为宣礼。礼拜前宣礼员在此处召唤穆斯林进寺礼拜。

Bangke Pavilion of Grand East Mosque

Located at Shanghexi Street, Xiaozhaokou, Jining City, it was built during the Hongwu period (1368-1368) of the Ming Dynasty. Bangke is the Persian pronunciation and means "calling to prayer." Here, Muezzin called Muslims to pray before worship.

漕井桥

　　大运河故道上的重要水利设施之一，位于济宁市安居镇桥西村。始建于清代，清顺治五年（1648年）、康熙十五年（1676年）重修。原为七孔，今残存两孔，基本完整。

Caojing Bridge

One of the important water conservancy facilities on the former grand canal, located at Qiaoxi Village, Anju Town, Jining City, it was built in the Qing Dynasty (1644-1911) with seven bridge openings. Now only two bridge openings remain.

荩园

位于任城区境内的戴庄花园内。戴庄原为明末清初著名画家戴鉴（号石坪）的别墅，后转给了当地官僚富豪李澍作为花园，改称为荩园。

Jin Garden

Located in Daizhuang Garden, Rencheng District, was the former house of famous painter Dai Jian of the late Ming Dynasty (1368-1644) and early Qing Dynasty (1644-1911). It was therefore named Daizhuang and renamed later as Jin Garden.

南阳古镇

位于微山县境内。京杭运河河畔的小镇，历史悠久。康熙皇帝和乾隆皇帝南巡时，曾在此停驻赏景。

Nanyang Ancient Town

Located in Weishan County was a small town beside the grand canal with a long history. Emperor Kangxi and Qianlong once stopped here to enjoy the view when they had their tour of south China.

清代古城门

位于微山县两城镇东南 7000 米，独山湖北岸。此古城门沉寂已久，见证着历史岁月的沧桑。

Ancient Town Gate of the Qing Qynasty

Located at 7000 meters to the southeast of Liangcheng Town, Weishan County and the north bank of Dushan Lake, this ancient town gate witnessed the vicissitudes of history silently for a long time.

魁星楼

位于金乡县，始建于明万历二十七年（1599年）。由汝南人彭鲲化任金乡知县而建。上层供奉魁星神像，下层明窗四敞。楼前架凌云桥于寿河之上，与陆岸相连。

Kuixing Tower

Located in Jinxiang County, it was built in 1599. There were Kuixing statues on the upper part and open windows on the lower part. A bridge called Lingyun Bridge was built across the Shou River in front of the tower, connecting it to the land.

> 冉子祠

现存三间硬山式建筑，建在 50 厘米高的石台上。坐北朝南，东西长九米，宽五米，高 5.5 米。重脊灰瓦，脊饰兽花纹。过冉子祠后门有冉子墓，呈圆形，祠前两棵古柏，据载这两棵树都是唐柏。

Ranzi Shrine

The existing 3 south-facing buildings were built on a 50-cm-high stone platform. They are 9 meters long from east to west, 5 meters wide and 5.5 meters high with double ridges decorated by animal patterns and grey tile. The round tomb of Ranzi is near the back door.

太白楼

位于济宁市城区古运河北岸，原是唐代贺兰氏经营的酒楼，因李白经常光顾而名声大振，生意兴隆。唐咸通二年（861年），吴兴人沈光敬慕李白，登楼观光，为其撰书"太白酒楼"匾额，并作《李翰林酒楼记》，从此贺兰氏酒楼以"太白楼"而闻名于世。

Taibai Restaurant

Located on the north bank of the ancient canal, Jining city, it was originally a restaurant run by Mrs. Helan in the Tang Dynasty (618-907). It became famous because of Li Bai's frequent visiting. Shen Guangjing wrote "Taibai Restaurant" for it on plaques in 861 and it was renamed after that.

宝相寺鼓楼

位于汶上县城西北隅。始建于盛唐，称昭空寺。北宋咸平五年（1003年）宋真宗禅封泰山时曾住跸于此。他看到整个寺院结构严谨，呈现富态宝相，遂改名为宝相寺。正面中路为山门，山门内左右分别为钟楼、鼓楼。

Baoxiang Temple Drum Tower

Located in the northwest of Wenshang County and built during the Tang Dynasty (618-907), the tower is located in what is known as Zhaokong Temple. In 1003, Emperor Song Zhenzong saw the temple had a rigorously made structure and an appearance that gave off signs of "good luck," so he renamed it Baoxiang Temple, which translates to "good luck appearance temple". The front gate is in the middle of the front road and has a bell tower to its left and a drum tower to its right.

宝相寺塔

又名太子灵踪塔，始建于北宋。塔高 45.5 米，砖砌八角形 13 层楼阁式建筑。

Baoxiang Temple Pagoda

Also known as Taizilingzong Pagoda, it was built in the Northern Song Dynasty (960-1127). It is a 45.5 meters high, 13-story pavilion-style brick pagoda shaped like an octagon.

九龙山摩崖造像石刻

位于曲阜市小雪街道武家村东约百米处,九龙山中部山体的西南山坡上。石刻刻于盛唐,造像共有大小石佛洞龛六处。

Jiulong Mountain Cliff Stone Statues

Located on the southwest slope of middle Jiulong Mountain, 100 meters to the east of Wujia Village, Xiaoxue Street, Qufu City, the stone statues were carved during the Tang Dynasty (618-907) and there were six cave shrines containing stone Buddhas in different sizes.

凤凰山唐代佛造像

鲁南凤凰山古名巨岳，山前有石窟，名大王窝，窟里有石像，高约四米，据史籍记载系唐代石刻。

Phoenix Mountain Buddha Statues in the Tang Dynasty

Formerly known as Juyue, the phoenix mountain in south Shandong has a stone cave called King Cave in front of it. There is a stone statue 4 meters high inside the cave. According to historical records, it was carved during the Tang Dynasty (618-907).

嘉祥武氏墓群石刻

位于嘉祥县纸纺镇武宅山村北，为汉代的祠堂和墓地。始建于东汉桓、灵时期（147~184年），石刻画像为"孔子见老子"。

Carved Stone of Mrs. Wu's Tombs in Jiaxiang

Located at north Wuzhaishan Village, Zhifang Town, Jiaxiang County, the tomb is located in what was an ancestral temple and cemetery in the Han Dynasty (25-220). Built in the Huan and Ling (147-184) periods, the carved stone shows a scene depicting the time Confucius met Laozi.

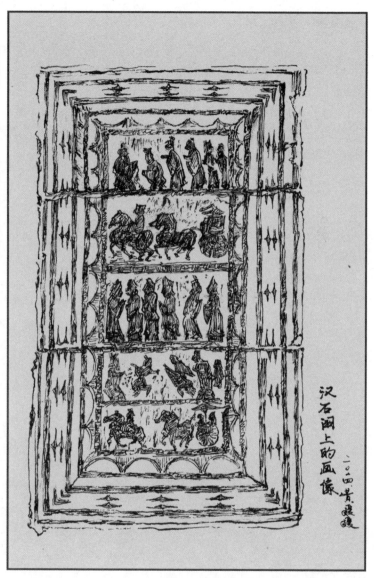

嘉祥武氏墓群石狮

　　原为武氏墓地神道设施。一对圆雕石狮相对立于阙前两侧，巨口膛目，昂首顾盼，浑朴端庄，形态生动。

Stone Lions of Mrs. Wu's Tombs in Jiaxiang
This was the original sacrifice facility of Mrs. Wu's tombs. A pair of circular engraved stone lions stand on both sides of the watchtower with open mouths, staring eyes, and a vivid, dignified appearance.

潘家大楼

原为军阀潘洪钧的私邸,位于古槐街道古槐路。平面分东、西、南、北四楼,回廊相接。砖石墙体,米浆灰灌缝,楠木柱檩,抬梁式木构架,硬山式屋顶。

Pan's Compound

Located at Guhuai Road, this was the former private residence of the warlord Pan Hongjun. There were buildings in the east, west, north and south respectively with winding corridors connecting to each other. The building had masonry walls, lift beam timber frames, and hard, mountain style roofs.

微山朝阳洞旁的民居

　　这些民居皆为石屋,散布于微山县两城乡独山岛上。石路、石屋在古树衬托下显得幽深、古朴而深邃。

Dwelling Houses beside Chaoyang Cave in Weishan

Situated on Dushan Island, Liangcheng Town, Weishan County, all the dwelling houses are stone houses. With the old trees as a backdrop, the stone path and stone houses looked quiet and simplistic.

> 日军碉堡

兖州现存有三座日军1938年修建的碉堡，两处在泗河西边的景观大道上，一处在九一医院门口。

The Japanese Fort

There are three Japanese forts built in 1938 that still exist in Yanzhou City. Two of them are located on Landscape Avenue, west of Sihe River, while one is located at the entrance of the 91 Hospital.

山头花鼓戏

在邹城市流行演唱的花鼓戏以大束镇山头村最为有名,演唱者自敲自唱,二人一组,历史可追溯到明嘉靖年间(1521~1567年)。

Shantou Flower Drum Opera

The flower drum opera of Shantou Village, Dashu Town, Zoucheng City, is the most famous of its kind. Performers sing while playing the drum in a group of 2 and it has a history of over 500 years.

兖州花棍舞

起源可上溯到宋元时期,"花棍"用一米多长竹竿制作,当中嵌几根铁轴,每轴穿三四枚铜制钱,舞时遍身拍打发出沙沙声响,棍两端饰以彩条、流苏,更增添了欢乐气氛。

Yanzhou Flower Stick Dancing

With a history of over 1000 years, the Flower Stick was made of a bamboo pole more than 1 meter long that was embedded in several iron axles made by 3 or 4 copper coins. The flower stick was decorated with color strips and tassels at both ends and rustled during the dancing. The result was a very joyful dance.

二人斗

古老的汉族民间舞蹈。由一个年轻艺人，身穿道具进行表演。道具是两个容貌怪异的"小鬼"，搂抱在一起打斗和摔跤的造型，他们或怒目圆睁，或怒发冲冠，或咬牙恨齿，皆表现为"斗气"的表情。

Two People Fighting

This style is an old folk dance of the Han people. This was performed by a young actor wearing props. The props are two odd-looking young boys who are huddled together to fight and wrestle in a "grudge" performance.

虎头袢子

渔家行船打鱼，使用两条长长的袢带把孩子拴在船的桅杆或船楼上，防止孩子不慎落水。袢带呈交叉形，两米多长。图上袢带的正面有一造型夸张的虎头。

Tiger Head Belt Loop

When sailing or fishing, fishermen use two long belt loops to tie children to the mast or deck house to prevent them from falling overboard. The drawing shows a belt loop with an exaggerated tiger head pattern.

嘉祥跑竹马

汉族民俗文化，舞者套进竹马中作骑马状，与搭档演出。其竹马是用竹篾和布扎成马（或驴）型骨架，分马头、马尾两截，中间是空的。

Riding Bamboo Horse in Jiaxiang

This comes from the folk culture of the Han people. The dancer rides a bamboo horse. The framework of the bamboo horse is made by bamboo skin and cotton and is divided into two parts: the head and the tail. The inside is hollow.

文圣拳

武术拳种之一，分武功、文功两部分。传说由老洪拳演化而来。"文"字取意于文功静坐之法，"圣"字显其拳理高深。主要在山东省嘉祥、汶上、微山、济宁以及江苏省沛县流传。

Wensheng Boxing

This kind of martial art was divided into two parts: Wu and Wen. "Wen" means to "sit still" and "Sheng", which refers to Wu, describes the profound theory of boxing. It's popular in Shandong and in Pei County of Jiangsu.

> 梁山梅花拳

中国传统武术中著名的拳种。内容丰富多彩，基本内容包括文理和武功两大类。

Liangshan Plum Blossom Boxing

This is a famous branch of traditional Chinese martial arts; it is rich in content, theory and action.

鱼台木版年画（一）

　　齐鲁民间艺术宝库中的奇葩，文化寓意深厚，有手绘、木版套印、木印填色三种。创始于唐代，在明清得以繁荣和发展，"家家点染，户户丹青"是鱼台木版年画繁盛的写照。

Yutai New Year Woodblock Pictures (1)

Yutai New Year Woodblock Pictures is a wonderful flower in folk art treasure house of Shandong. (check for accuracy of translation) It has profound cultural implications and contains a hand-painted, woodblock overprint and three types of woodprint coloring. Originally from the Tang Dynasty (618-907) and developed in the Ming (1368-1644) and Qing (1644-1911) dynasties, it was popular and highly praised by thousands of families.

133

鱼台木版年画（二）

以线条粗犷、色彩绚丽、造型简练、构图饱满、装饰性强为特色。装饰手法以丹作底色，色彩强烈富丽。人物衣饰上的花纹用线条流畅的写金描银渲染，俗称"写花"，极富民间画的韵味，为其他地区民间年画所罕见。

Yutai New Year Woodblock Pictures (2)

With bold lines, bright colors, simple modeling, and full composition and decoration, the pictures are extremely rich examples of this folk painting style and are a rarity compared with the folk pictures of other regions.

临沂市
Linyi City

> 宝泉寺

位于罗庄区朱陈村西涑河之畔。始建于宋，近代毁于战火。现仅存墓塔数座，石碑数块。图为宝泉寺塔林。

Baoquan Temple
Built in the Song Dynasty (960-1127), the temple was destroyed by war in modern times. It is located beside Xisu River, Zhuchen Village, Luozhuang district and only a few tomb towers and stele remnants remain today.

迎仙桥

位于蒙阴县城东 28 千米处的旧寨乡北楼村西南角,始建于明初,清光绪三十三年（1907 年）重修。

Yingxian Bridge

Located at the southwest corner of Beilou Village, Jiuzhai Town, 28 kilometers to the east of Mengyin County, the bridge was built during the early Ming Dynasty (1368-1644) and rebuilt in 1907.

蒙山千年古村落

位于蒙山主峰东麓，群山环抱，环境优美。一座千年古村落，一片百年石头房，这里是沂蒙山人最早居住的古村落。

1000-Year-Old Ancient Village in Mengshan

Located at the eastern foot of Mengshan Mountain and surrounded by mountains with beautiful scenery, this 1000-year-old ancient village and 100-year-old stone houses were the ancient dwellings of the Yimeng mountain people.

马牧池乡村落

　　位于沂蒙山区的核心地带，至今仍保留着古朴的村落、幽深的石板巷、石板桥以及残破的老石屋、茅草屋。

Village in Mamuchi Town

Located in the heartland of the Yimeng mountain area, the town still retains the primitive village, a quiet slate lane, an old stone bridge, a wrecked stone house, and a thatched house.

【临沂五贤祠】

　　原位于老城南关外名景贤祠，明嘉靖年间（1522~1566年）沂州知州何格所建，祭祀诸葛亮、王祥、王览、颜真卿、颜杲卿五人。大殿前是御碑亭，亭内有乾隆题诗碑，其诗为："孝能竭力王祥贤，忠以捐躯颜杲真。所遇由来殊出处，端推诸葛是全人。"

Linyi Wuxian Shrine
Originally located at the Mingjingxian Shrine of old town, it was built by Yizhou official He Ge from 1522 to 1566 to worship the five famous sages of history. There was a royal pavilion with an inscription written by Emperor Qianlong in front of the grand hall.

吴白庄汉墓

位于临沂市罗庄区盛庄街道吴白庄村，最早公开发掘于1972年。墓葬为半地上砖石结构建筑，此图是石墓前室中过梁上半月形门额（对鸟）。

Tomb of Han Dynasty in Wubaizhuang Village

Located at Wubaizhuang Village, Shengzhuang Street, Luozhuang District, Linyi City and excavated in 1972, the tomb was a half-overgrown brick building. The drawing shows two birds on the half-moon-shaped lintel of the front room of the tomb.

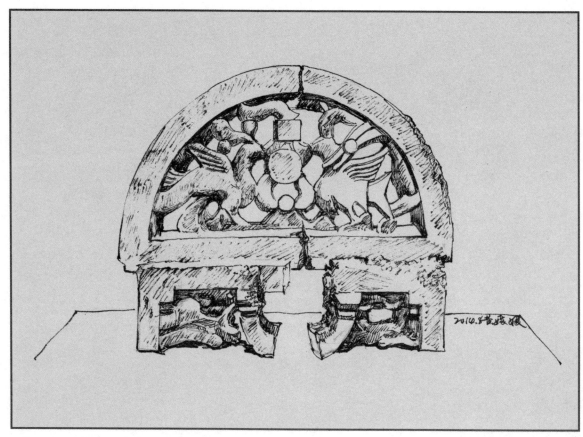

> 丛柏庵

　　位于费县许家崖，是临沂市唯一的尼姑庵。始建于隋代，重建于明嘉靖三十九年（1560年），以侧柏密集而得名。

Cypress Nunnery

Located at Xujiaya, Fei County, this is the only nunnery in Linyi City. Built in the Sui Dynasty (581-618), it was rebuilt in 1560 and named after the dense cypress trees that surround it.

地主大院

沂南县马牧池乡常山庄地主李忠奉家的大院,典型鲁南传统民居,由众多房屋组成。

Landlord Compound

The compound of landlord Li Zhongfeng of Changshan Village, Mamuchi Town, Yinan County consists of numerous houses.

> 戏台子

　　沂南县马牧池乡常山庄的戏台子,原是村里老少看戏娱乐的场所,现已废弃。

Opera Stage

This was the former opera stage of Changshan Village, Mamuchi Town, Yinan County. It was an entertainment venues for villagers before and is now abandoned.

马牧池民居

  沂南县马牧池乡的民居，山清水秀、民风浑厚。村居简练古朴，茅舍土墙，处处展现着村落的历史文化。

Mamuchi Dwelling Houses

The dwelling houses of Mamuchi Town, Yinan County, with picturesque scenery and honest people, capture the primitive simplicity of the time. The cottages and earthen walls show the history and culture of the village.

王庄乡天主教堂

位于沂水县夏蔚镇王庄村,建于清光绪十九年(1893年),原有房屋百余间。可容三四百人,该教堂为掩护山东省委、八路军及《大众日报》创刊做出重大贡献。

Wangzhuang Cathedral

Located at Wangzhuang Village, Xiawei Town, it was built in 1893 with over 100 rooms as the village cathedral. The cathedral is capable of hosting 300 to 400 people.

临沂天主教堂

位于临沂市区兰山路中段,是山东省唯一的古罗马式大教堂,是临沂城古建筑当中比较古老的建筑之一。有拱形圆顶、大型石柱、雕花柱头等。

Linyi Cathedral

Located at middle Lanshan Road, Linyi City, this is the only Roman style cathedral in Shandong and is the center of the Linyi Catholic community. An arched dome, giant stone column and carving column head are features of this cathedral.

| 华东革命烈士陵园 |

位于市区东南部沂河西岸金雀、银雀二山之间的平地上。1949年为纪念在抗日战争和解放战争中牺牲的华东地区革命烈士而建，是华东地区最大的革命烈士陵园。图为革命烈士纪念塔，五角灯塔式建筑，高47.5米。

Revolutionary Martyr Cemetery of East China

Located on the flat ground between the Jinque and Yinque mountains on the west bank of Yihe River, southeast of Linyi City, it was built in April, 1949 by People's Government of Shandong Province to commemorate the revolutionary martyrs who sacrificed themselves in the war against Japan and the war for liberation.

银雀山汉墓竹简博物馆

位于临沂市兰山区沂蒙路212号。是我国第一座遗址性专题汉墓竹简博物馆,为古典宫廷式建筑。

Yinqueshan Bamboo Slips Museum

Located at No.212 Yimeng Road, Lanshan District, Linyi City, the museum was built in classic palace style architecture and is the first on-site museum for bamboo slips from tombs of the Han Dynasty (25-220).

> 中共山东分局旧址

　　位于沂水县王庄乡，1938~1939 年中共山东分局在此办公。期间八路军山东纵队在此成立，于 1939 年 1 月 1 日创刊《大众日报》，罗荣桓、徐向前等元帅在此领导指挥山东抗战。

Site of the Communist Party of China Committee, Shandong Branch

Located at Wangzhuang Village, Yishui County, it was the site of the Shandong branch of the Communist party of China from 1938 to 1939. On January 1, 1939, the Public Daily started its publication here.

151

西汉陶车陶马俑

　　临沂银雀山八号墓中出土，尤其是马儿体型肥壮，人物生动，展现了西汉时期精湛的制陶技术。

Pottery Cart and Pottery Horse of West Han Dynasty

Unearthed from the No. 8 tomb in Yinqueshan mountain, Linyi City, the stout and strong body of the horse vividly shows the exquisite ceramics technology of the West Han Dynasty (202 B.C.-25 A.D.).

[西汉陶俑]

临沂银雀山八号墓中出土。两手相抱置于腹际,姿势贴切自然。面庞清秀俊雅,表情温顺娴雅,体态端庄大方,体态匀称协调,

Pottery Figures of West Han Dynasty

Unearthed from No. 8 tomb in Yinque mountain, Linyi City, the figure has a delicate and pretty face, gentle and refined expression, dignified posture and is displayed holding two hands on its belly.

153

小郭泥塑（一）

相传起源于清咸丰年间（1851~1861年），以兰陵县兴明乡（现属向城镇）小郭村为代表。传统手工捏制的"小郭泥人"已有近200年历史。泥塑形象生动传神、异彩纷呈。

Xiaoguo Clay Sculpture (1)

According to legend, it originated in the Xianfeng period (1851-1861) of the Qing Dynasty, represented by Xiaoguo Village. The traditional handmade Xiaoguo clay sculpture has a history of nearly 200 years. The clay images are vivid and colorful.

小郭泥塑（二）

制作过程比较复杂，费时较长，和泥、做坯、晾晒、描彩等。泥塑的形象取材广泛，有戏曲人物、畜禽鸟兽等。

Xiaoguo Clay Sculpture (2)

The process of making clay figurines is complicated and time-consuming; it includes mixing the mud, making preforms drying the clay, coloring it, and so on. Clay sculptures have various subjects, such as opera figures, animals and birds.

布老虎

　　制作布老虎的风俗有悠久的历史，但是具体起源于何时已不可考。明代就很兴盛，清代后期达到高潮。虎在临沂民间是勇猛、吉祥、安全的象征。

Cloth Tiger

The cloth tiger has a long history, but the exact time of origin cannot be determined. It was popular in the Ming (1368-1644) and Qing (1644-1911) dynasties and reached its heyday in the Qing Dynasty. The tiger is the symbol of bravery, good look and safety.

送火神之踩高跷

沂水县诸葛镇大峪村的送火神的祭祀活动由来已久,据说始于清朝,已有 500 多年的历史,每年的农历正月初五举行。图中表现的是一位穿着老戏袍踩高跷者,正准备参与出行。

Walking on Stilts to Send the God of Fire

This is a worship ceremony of the Dayu Village, Zhuge Town, Yishui County. With a history of over 500 years, it is held on January 5th of the lunar calendar every year. The drawing shows somebody wearing an old opera gown walking on stilts, who appears to be ready to go.

弦子戏

　　古老剧种，以抒情、叙事见长的唱工戏。拥有几百支传统音乐曲牌，唱腔委婉、细腻、缠绵。现流行于沂南县。

Xianzi Opera

Popular in Yinan County, it is an old style of opera with hundreds of branches of traditional music. It is good at expressing emotion and narrating with a soft, delicate and tender voice.

绣香包

农历五月初五是中国传统的端午节，民间有"戴香包求健康"的传统习俗。这些"香包"一般用彩色丝布做成，内有朱砂、雄黄等散发浓郁芳香的中药材料。

Sachet

May 5th by the lunar calendar is the day for the traditional Chinese Dragon Boat Festival. People wear sachets on this day to wish for health as per folk customs. These sachets are generally made of colorful silk cloth and have traditional Chinese medicine in them that produces rich aromas, like cinnabar and realgar.